100 facts on
The Human Body

100 facts on

The Human Body

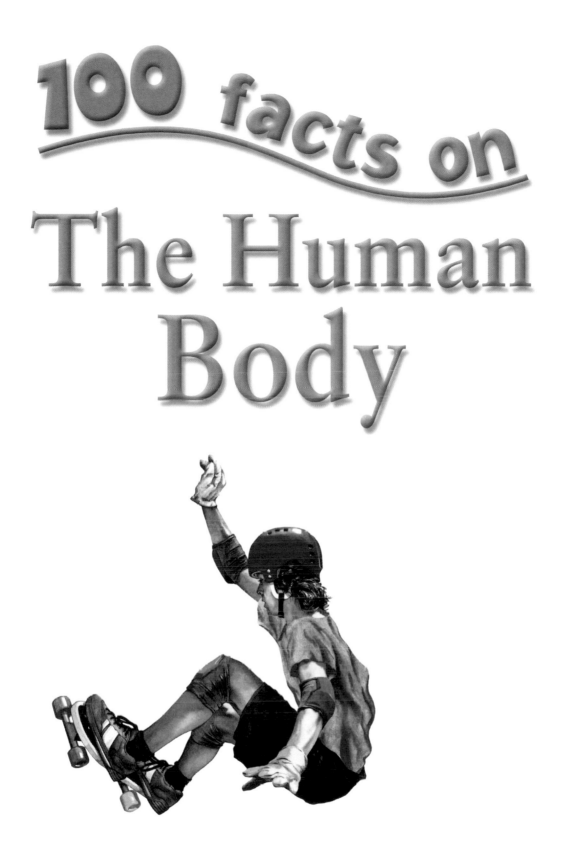

Steve Parker

Consultant: Dr. Kristina Routh MB ChB

First published as hardback in 2003 by Miles Kelly Publishing Ltd
Harding's Barn, Bardfield End Green, Thaxted, Essex, CM6 3PX, UK

This edition printed 2010

2 4 6 8 10 9 7 5 3 1

Editorial Director: Belinda Gallagher
Art Director: Jo Brewer
Assistant Editor: Lucy Dowling
Volume Designer: John Christopher, White Design
Copy Editor: Sarah Ridley
Proofreader: Hayley Kerr
Indexer: Jane Parker
Production Manager: Elizabeth Collins
Reprographics: Anthony Cambray, Liberty Newton, Ian Paulyn
Assets Manager: Bethan Ellish

ISBN 978-1-84236-765-0

Printed in China

British Library Cataloguing-in-Publication Data
A catalogue record for this book is available from the British Library

ACKNOWLEDGEMENTS
The publishers would like to thank the following artists
who have contributed to this book:
Syd Brak, Mike Foster/Maltings Partnership, Janos Marffy,
Martin Sanders, Mike Saunders, Rudi Vizi

Cartoons by Mark Davis at Mackerel

All other images from the Miles Kelly Archives

Made with paper from a sustainable forest

www.mileskelly.net
info@mileskelly.net

www.factsforprojects.com

Contents

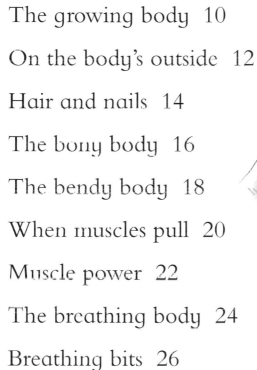

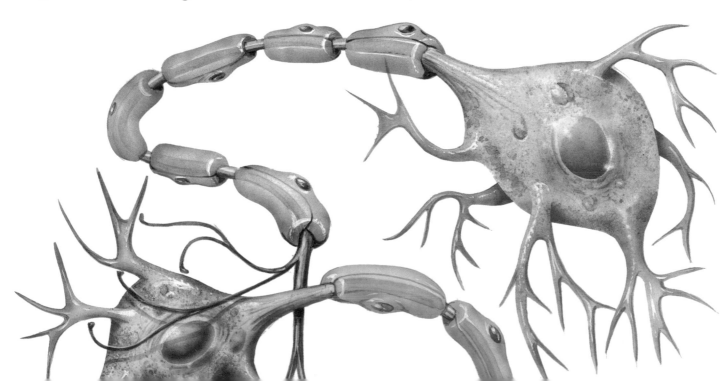

Outside, inside

1 **There are more than six billion human bodies in the world.** If you could say hello to all of them, even quickly, it would take you more than 300 years. In some ways, all of these human bodies are very similar, especially on the inside. Each one has a heart and brain, bones and guts, arms and legs and skin. But each human body is also individual, especially on the outside. You have your own appearance, size and shape, facial features, hairstyle and clothes. You also have your own personality, with likes and dislikes, and special things that make you happy or sad. So human bodies may be very similar in how they look, but not in what they do. You are unique, your own self.

▶ We tend to notice small differences on the outside of human bodies, such as height, width, hair colour and clothes. This allows us to recognize our family and friends.

Baby body

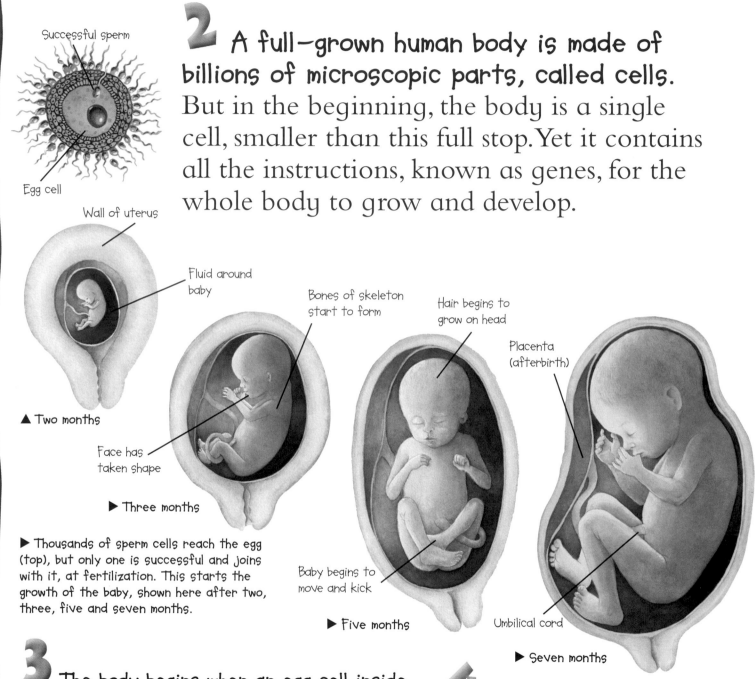

Successful sperm

Egg cell

Wall of uterus

Fluid around baby

▲ Two months

Bones of skeleton start to form

Face has taken shape

▶ Three months

Hair begins to grow on head

Placenta (afterbirth)

Baby begins to move and kick

▶ Five months

Umbilical cord

▶ Seven months

2 A full-grown human body is made of billions of microscopic parts, called cells. But in the beginning, the body is a single cell, smaller than this full stop. Yet it contains all the instructions, known as genes, for the whole body to grow and develop.

▶ Thousands of sperm cells reach the egg (top), but only one is successful and joins with it, at fertilization. This starts the growth of the baby, shown here after two, three, five and seven months.

3 The body begins when an egg cell inside the mother joins up with sperm from the father. The egg cell splits into two cells, then into four cells, then eight, and so on. The bundle of cells embeds itself in the mother's womb (uterus), which protects and nourishes it. Soon there are thousands of cells, then millions, forming a tiny embryo. After two months the embryo has grown into a tiny baby, as big as your thumb, with arms, legs, eyes, ears and mouth.

4 After nine months in the womb, the baby is ready to be born. Strong muscles in the walls of the womb tighten, or contract. They push the baby through the opening, or neck of the womb, called the cervix, and along the birth canal. The baby enters the outside world.

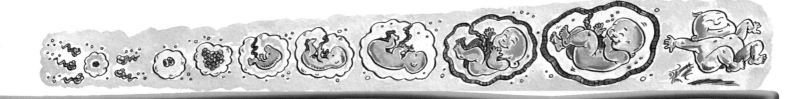

5 A newborn baby may be frightened and usually starts to cry. Inside the womb it was warm, wet, dark, quiet and cramped. Outside there are lights, noises, voices, fresh air and room to stretch. The crying is also helpful to start the baby breathing, using its own lungs.

▼ Nine months

Wall of womb is stretched

Placenta

◄ Inside the womb, the baby cannot breathe air or eat food. Nutrients and oxygen pass from mother to baby through the blood vessels in the ropelike umbilical cord.

Umbilical cord

6 Being born can take an hour or two – or a whole day or two. It is very tiring for both the baby and its mother. After birth, the baby starts to feel hungry and it feeds on its mother's milk. Finally, mother and baby settle down for a rest and some sleep.

Baby is born head-first

Cervix (neck of womb)

The growing body

7 A new baby just seems to eat, sleep and cry. It feeds on milk when hungry and sleeps when tired. Also, it cries when it is too hot, too cold, or when its nappy needs changing.

8 A new baby is not totally helpless. It can do simple actions called reflexes, to help it survive. If something touches the baby's cheek, it turns its head to that side and tries to suck. If the baby hears a loud noise, it opens its eyes wide, throws out its arms and cries for help. If something touches the baby's hand and fingers, it grasps tightly.

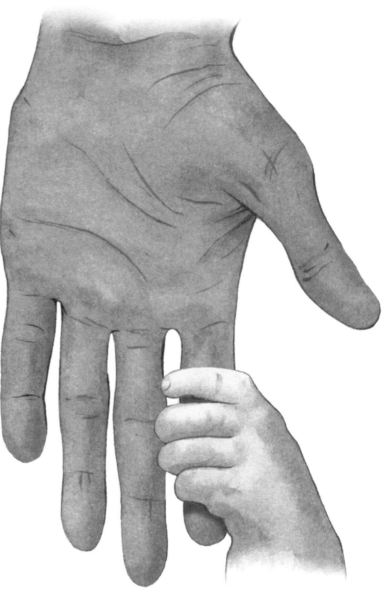

▲ In the grasping reflex, the baby tightly holds anything that touches its hand or fingers. Its grip is surprisingly strong!

WHAT HAPPENS WHEN?

Most babies learn to do certain actions in the same order. The order is mixed up here. Can you put it right?

walk, crawl, roll over, sit up, smile, stand

Answers:
smile, roll over, sit up, crawl, stand, walk

9 A new baby looks, listens, touches and quickly learns. Gradually it starts to recognize voices, faces and places. After about six weeks, it begins to smile. Inside the body, the baby's brain is learning very quickly. The baby soon knows that if it laughs, people will laugh back. If it cries, someone will come to look after it.

▼ Most babies crawl before they walk, but some go straight from sitting or 'bottom-shuffling' to walking.

11 As a baby grows into a child, at around 18 months, it learns ten new words every day, from 'cat' and 'dog' to 'sun' and 'moon'. There are new games such as piling up bricks, new actions such as throwing and kicking, and new skills such as using a spoon at mealtimes and scribbling on paper.

10 At about three months old, most babies can reach out to hold something, and roll over when lying down. By the age of six months, most babies can sit up and hold food in their fingers. At nine months, many babies are crawling well and perhaps standing up. By their first birthday, many babies are learning to walk and starting to talk.

12 At the age of five, when most children start school, they continue to learn an amazing amount. This includes thinking or mental skills such as counting and reading, and precise movements such as writing and drawing. They learn out of the classroom too – how to play with friends and share.

▶ Playing is lots of fun, but it's learning too, as children develop control over the muscles in their fast-growing bodies.

On the body's outside

13 Skin's surface is made of tiny cells which have filled up with a hard, tough substance called keratin, and then died. So when you look at a human body, most of what you see is 'dead'! The cells get rubbed off as you move, have a wash and get dry.

▲ Skin may feel smooth, but its surface is made of millions of tiny flakes, far too small to see.

14 Skin rubs off all the time, and grows all the time too. Just under the surface, living cells make more new cells that gradually fill with keratin, die and move up to the surface. It takes about four weeks from a new skin cell being made to when it reaches the surface and is rubbed off. This upper layer of skin is called the epidermis.

▼ This view shows skin magnified (enlarged) about 50 times.

15 Skin's lower layer, the dermis, is thicker than the epidermis. It is made of tiny, bendy, threadlike fibres of the substance collagen. The dermis also contains small blood vessels, tiny sweat glands, and micro-sensors that detect touch.

Hair

Oil gland

Pain sensors

Light touch sensor

Hair follicle

Epidermis

Dermis

Heavy pressure sensor

▼ Skin is tough, but it sometimes needs help to protect the body. Otherwise it, and the body parts beneath, may get damaged.

Safety helmet protects head and brain

Elbow-pads cushion fall

Knee-pads prevent hard bumps

Gloves save fingers from scrapes and breaks

16 One of skin's important jobs is to protect the body. It stops the delicate inner parts from being rubbed, knocked or scraped. Skin also prevents body fluids from leaking away and it keeps out dirt and germs.

17 Skin helps to keep the body at the same temperature. If you become too hot, sweat oozes onto your skin and, as it dries, draws heat from the body. Also, the blood vessels in the lower layer of skin widen, to lose more heat through the skin. This is why a hot person looks sweaty and red in the face.

18 Skin gives us our sense of touch. Millions of microscopic sensors in the lower layer of skin, the dermis, are joined by nerves to the brain. Different sensors detect different kinds of touch, from a light stroke to heavy pressure, heat or cold, and movement. Pain sensors detect when skin is damaged. Ouch!

SENSITIVE SKIN
You will need:
a friend   sticky-tack
two used matchsticks   ruler
1. Press some sticky-tack on the end of the ruler. Press two matchsticks into the sticky-tack, standing upright, about 1 centimetre apart.
2. Make your friend look away. Touch the back of their hand with both matchstick ends. Ask your friend: 'Is that one matchstick or two?' Sensitive skin can detect both ends.
3. Try this at several places, such as on the finger, wrist, forearm, neck and cheek.

Hair and nails

19 There are about 120,000 hairs on the head, called scalp hairs. There are also eyebrow hairs and eyelash hairs. Grown-ups have hairs in the armpits and between the legs, and men have hairs on the face. And everyone, even a baby, has tiny hairs all over the body – 20 million of them!

▼ Hair contains pigments (coloured substances) – mainly melanin (dark brown) and some carotene (yellowish). Different amounts of pigments, and the way their tiny particles are spread out, cause different hair colours.

▼ Black curly hair is the result of black melanin from a flat hair follicle

▲ Straight red hair is the result of red melanin from a round hair follicle

▲ Blonde wavy hair is the result of carotene from an oval hair follicle

20 Each hair grows from a deep pit in the skin, called a follicle. The hair is only alive where it gets longer, at its base or root, in the bottom of the follicle. The rest of the hair, called the shaft, is like the surface of the skin – hard, tough, dead and made of keratin. Hair helps to protect the body, especially where it is thicker and longer on the head. It also helps to keep the body warm in cold conditions.

▶ Straight black hair is the result of black melanin from a round follicle

21 Scalp hairs get longer by about 3 millimetres each week, on average. Eyebrow hairs grow more slowly. No hairs live for ever. Each one grows for a time, then it falls out, and its follicle has a 'rest' before a new hair sprouts. This is happening all the time, so the body always has some hairs on each part.

22 Nails, like hairs, grow at their base (the nail root) and are made of keratin. Also like hairs, nails grow faster in summer than in winter, and faster at night than by day. Nails lengthen by about half a millimetre, on average, each week.

▼ The growing nail root is hidden under skin. The nail slides slowly along the nail bed.

Nail root

Cuticle (skin edge)

Nail bed

Bone inside finger

23 Nails have many uses, from peeling off sticky labels to plucking guitar strings or scratching an itch. They protect and stiffen the ends of the fingers, where there are nerves that give us our sense of touch.

► Nails make the fingertips stronger and more rigid for pressing hard on guitar strings. Slightly longer nails pluck the strings.

I DON'T BELIEVE IT!

A scalp hair grows for up to five years before it falls out and gets replaced. Left uncut during this time, it would be about 1 metre long. But some people have unusual hair that grows faster and for longer. Each hair can reach more than 5 metres in length before dropping out.

The bony body

24 The human body is strengthened, supported and held up by parts that we cannot see – bones. Without bones, the body would be as floppy as a jellyfish! Bones do many jobs. The long bones in the arms work like levers to reach out the hands. The finger bones grasp and grip. The leg bones are also levers when we walk and run. Bones protect softer body parts. The domelike skull protects the brain. The ribs in the chest are like the bars of a cage to protect the heart and lungs inside. Bones also produce blood cells, as explained on the opposite page.

▶ The skeleton forms a strong framework inside the body. The only artificial (man-made) substances that can match bone for strength and lightness are some of the materials used to make racing cars and jet planes.

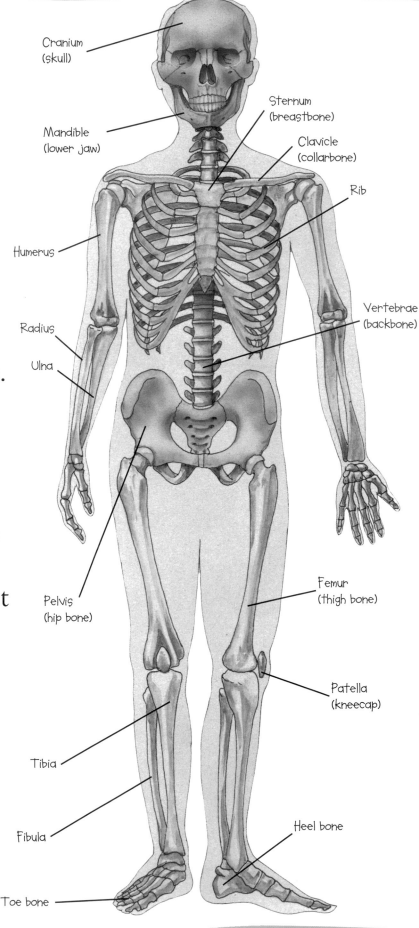

Cranium (skull)

Mandible (lower jaw)

Sternum (breastbone)

Clavicle (collarbone)

Rib

Humerus

Radius

Ulna

Vertebrae (backbone)

Pelvis (hip bone)

Femur (thigh bone)

Patella (kneecap)

Tibia

Fibula

Heel bone

Toe bone

25 All the bones together make up the skeleton. Most people have 206 bones, from head to toe as follows:

- 8 in the upper part of the skull, the cranium or braincase
- 14 in the face
- 6 tiny ear bones, 3 deep in each ear
- 1 in the neck, which is floating and not directly connected to any other bone
- 26 in the spinal column or backbone
- 25 in the chest, being 24 ribs and the breastbone
- 32 in each arm, from shoulder to fingertips (8 in each wrist)
- 31 in each leg, from hip to toetips (7 in each ankle)

◀ The skull has deep bowls for the eyes, and small holes where nerves pass through to join the brain inside.

NAME THE BONE!

Every bone has a scientific or medical name, and many have ordinary names too. Can you match up these ordinary and scientific names for various bones?

1. Mandible 2. Femur 3. Clavicle
4. Pelvis 5. Patella 6. Sternum

a. Thigh bone b. Breastbone
c. Kneecap d. Hip bone
e. Collarbone f. Lower jaw bone

Answers:
1f 2a 3e 4d 5c 6b

▶ Bone has a hard layer outside, a spongy layer next, and soft marrow in the middle.

26 Bone contains threads of the tough, slightly bendy substance called collagen. It also has hard minerals such as calcium and phosphate. Together, the collagen and minerals make a bone strong and rigid, yet able to bend slightly under stress. Bones have blood vessels for nourishment and nerves to feel pressure and pain. Also, some bones are not solid. They contain a jellylike substance called marrow. This makes tiny parts for the blood, called red and white blood cells.

Marrow

Spongy bone

Compact (hard) bone

End or head of bone

Nerves and blood vessels

'Skin' of bone (periosteum)

The bendy body

27 Without joints, almost the only parts of your body that could move would be your tongue and eyebrows! Joints between bones allow the skeleton to bend. You have more than 150 joints. The largest are in the hips and knees. The smallest are in the fingers, toes, and between the tiny bones inside each ear which help you hear.

28 There are several kinds of joints, depending on the shapes of the bone ends, and how much the bones can move. Bend your knee and your lower leg moves forwards and backwards, but not sideways. This is a hinge-type joint. Bend your hip and your leg can move forwards, backwards, and also from side to side. This is a ball-and-socket joint.

Collarbone

▶ In the shoulder, the upper arm bone's rounded head fits into a socket in the shoulder blade.

Head of upper arm bone

Shoulder blade

TEST YOUR JOINTS

Try using these different joints carefully, and see how much movement they allow. Can you guess the type of joint used in each one – hinge or ball-and-socket?

1. Fingertip joint (smallest knuckle)
2. Elbow
3. Hip
4. Shoulder

Answers:
1. hinge 2. hinge
3. ball-and-socket 4. ball-and-socket

29 Inside a joint where the bones come together, each bone end is covered with a smooth, shiny, slippery, slightly springy substance, known as cartilage. This is smeared with a thick liquid called synovial fluid. The fluid works like the oil in a car, to smooth the movements and reduce rubbing and wear between the cartilage surfaces.

30 The bones in a joint are linked together by a baglike part, the capsule, and strong, stretchy, straplike ligaments. The ligaments let the bones move but stop them coming apart or moving too far. The shoulder has seven strong ligaments.

◀ The arm joints are very flexible, but they can also work as strongly as the leg joints to hold up the whole body.

31 In some joints, there are cartilage coverings over the bone ends and also pads of cartilage between the cartilage! These extra pads are called articular discs. There is one in each joint in the backbone, between the spinal bones, which are called vertebrae. There are also two of these extra cartilages, known as menisci, in each knee joint. They help the knee to 'lock' straight so that we can stand up without too much effort.

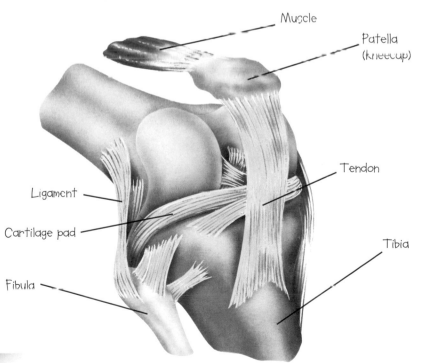

Muscle

Patella (kneecap)

Tendon

Ligament

Cartilage pad

Tibia

Fibula

▲ The knee has many ligaments, cartilage pads (menisci) and strong tendons that anchor muscles.

When muscles pull

32 Almost half the body's weight is muscles, and there are more than 640 of them! Muscles have one simple but important job, which is to get shorter, or contract. A muscle cannot get longer.

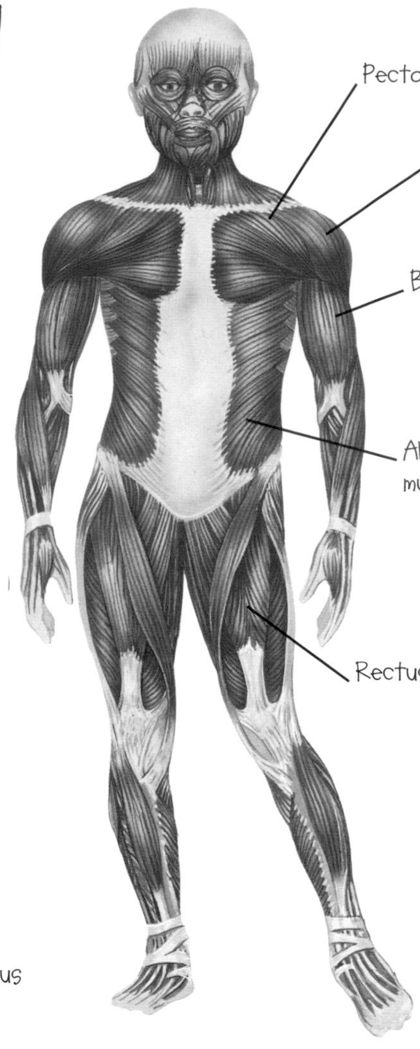

33 A muscle is joined to a bone by its tendon. This is where the end of the muscle becomes slimmer or tapers, and is strengthened by strong, thick fibres of collagen. The fibres are fixed firmly into the surface of the bone.

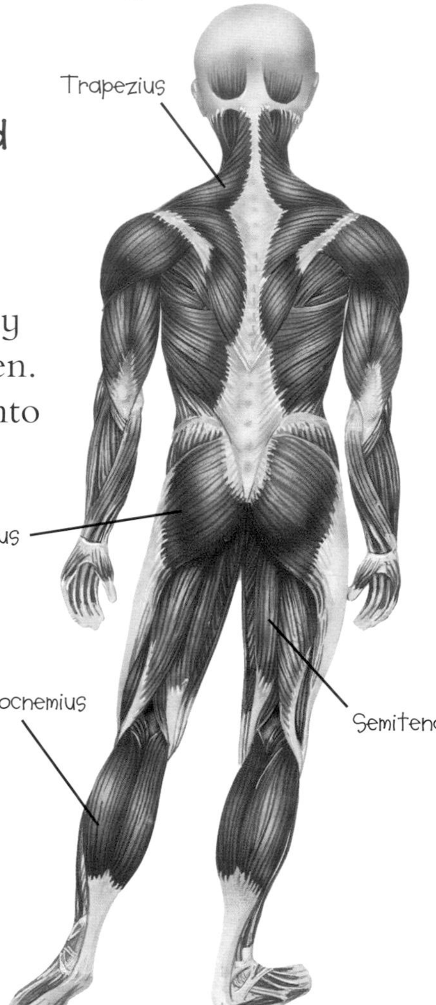

▼ A tendon is stuck firmly into the bone it pulls, with a joint stronger than superglue!

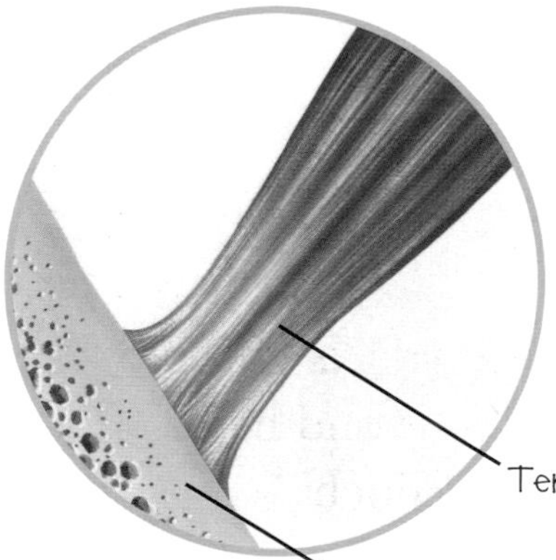

Tendon

Bone

Trapezius

Gluteus

Gastrocnemius

Semitendinosus

Pectoralis

Deltoid

Biceps

Abdominal wall muscles

Rectus femoris

▲ The muscles shown here are those just beneath the skin, called superficial muscles. Under them is another layer, the deep muscle layer. In some areas there is an additional layer, the medial muscles.

34 Some muscles are wide or broad, and shaped more like flat sheets or triangles. These include the three layers of muscles in the lower front and sides of the body, called the abdominal wall muscles. If you tense or contract them, they pull your tummy in to make you look thinner.

35 Most muscles are long and slim, and joined to bones at each end. As they contract they pull on the bones and move them. As this happens, the muscle becomes wider, or more bulging in the middle. To move the bone back again, a muscle on the other side of it contracts, while the first muscle relaxes and is pulled longer.

◀ A weightlifter's muscles can raise more than three times the body weight above the head.

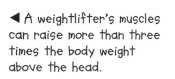

36 Every muscle in the body has a scientific or medical name, which is often quite long and complicated. Some of these names are familiar to people who do exercise and sports. The 'pecs' are the pectoralis major muscles across the chest. The 'biceps' are the biceps brachii muscles in the upper arms, which bulge when you bend your elbow.

37 If you take plenty of exercise or play sport, you do not gain new muscles. But the muscles you have become larger and stronger. This keeps them fit and healthy. Muscles which are not used much may become weak and floppy.

▶ Muscles work in two-way pairs, like the biceps and triceps, which bend and straighten the elbow.

Biceps

Triceps

Biceps gets shorter and the elbow moves

To move the arm back down, the triceps shortens and the biceps gets longer

21

Muscle power

38 **Muscles have many shapes and sizes, but inside they are all similar.** They have bundles of long, hairlike threads called muscle fibres, or myofibres. Each muscle fibre is slightly thinner than a hair. A big muscle has many thousands of them. Most are about 3 or 4 centimetres long. In a big muscle, many fibres of different lengths lie alongside each other and end-to-end.

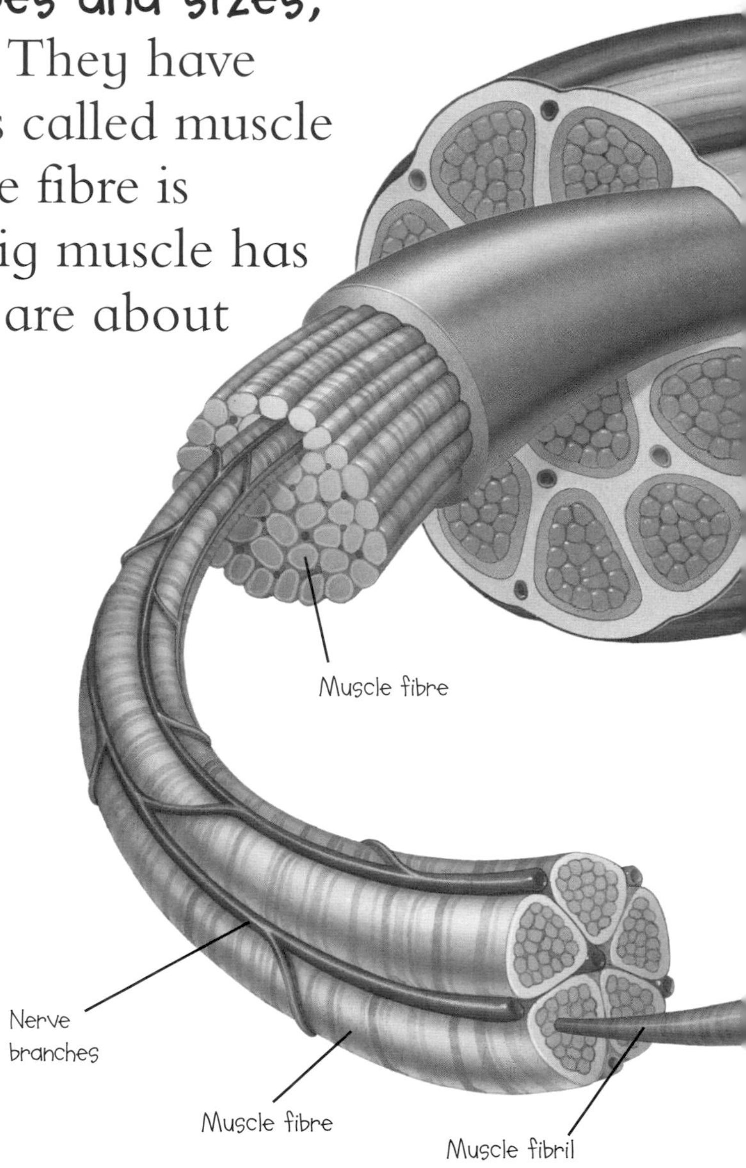

Muscle fibre

Nerve branches

Muscle fibre

Muscle fibril

▶ While arm muscles prepare to make the racket hit the ball, hundreds of other muscles keep the body poised and balanced.

39 **Each muscle fibre is made of dozens or hundreds of even thinner parts, called muscle fibrils or myofibrils.** There are millions of these in a large muscle. And, as you may guess, each fibril contains hundreds of yet thinner threads! There are two kinds, actin and myosin. As the actins slide past and between the myosins, the threads get shorter – and the muscle contracts.

40

Muscles are controlled by the brain, which sends messages to them along stringlike nerves. When a muscle contracts for a long time, its fibres 'take turns'. Some of them shorten powerfully while others relax, then the contracted ones relax while others shorten, and so on.

◄ The main part of a muscle is the body or belly, with hundreds of muscle fibres inside.

Body of muscle

Actin

Myosin

▼ Dozens of arm and hand muscles move a pen precisely, a tiny amount each time.

WHICH MUSCLES?

Can you match the names of these muscles, with different parts of the body?

a. Gluteus maximus b. Masseter
c. Sartorius d. Cardiac muscle
e. Pectoralis major

1. Heart 2. Chest 3. Front of thigh
4. Buttock 5. Mouth

Answers:
a4 b5 c3 d1 e2

41

The body's biggest muscles are the ones you sit on – the gluteus maximus muscles in the buttocks. The longest muscle is the sartorius, across the front of the thigh. Some of its fibres are more than 30 centimetres in length. The most powerful muscle, for its size, is the masseter in the lower cheek, which closes the jaws when you chew.

The breathing body

42 The body cannot survive more than a minute or two without breathing. This action is so important, we do it all the time without thinking. We breathe to take air into the body. Air contains the gas oxygen, which is needed to get energy from food to power all of the body's vital life processes.

▶ Body parts make up the respiratory system in the head, neck and chest. These carry out the process of breathing air to take oxygen into the body.

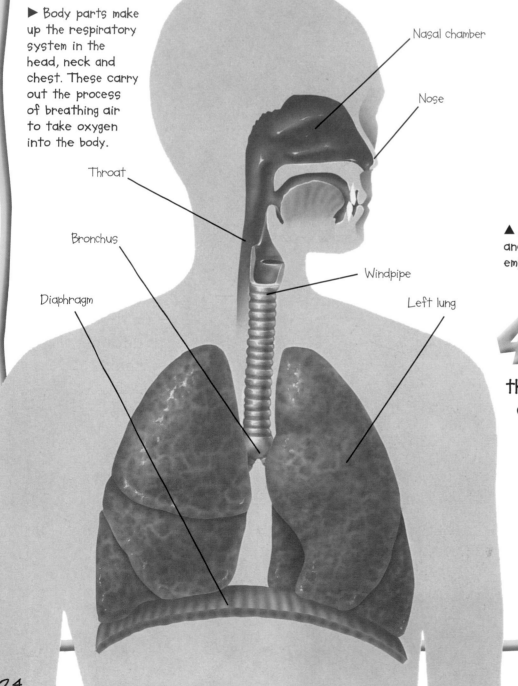

Nasal chamber

Nose

Throat

Bronchus

Diaphragm

Windpipe

Left lung

▲ A space suit protects an astronaut, and contains air to breathe in the emptiness of space.

43 Parts of the body that work together to carry out a main task are called a system — so the parts that carry out breathing are the respiratory system. These parts are the nose, throat, windpipe, the air tubes or bronchi in the chest, and the lungs.

44 **The nose is the entrance for fresh air to the lungs – and the exit for stale air from the lungs.** The soft, moist lining inside the nose makes air warmer and damper, which is better for the lungs. Tiny bits of floating dust and germs stick to the lining or the hairs in the nose, making the air cleaner.

45 **The windpipe, or trachea, is a tube leading from the back of the nose and mouth, down to the lungs.** It has about 20 C shaped hoops of cartilage in its wall to keep it open, like a vacuum cleaner hose. Otherwise the pressure of body parts in the neck and chest would squash it shut.

▲ The human voice can make a wide range of sounds, from loud to soft, and low to high.

46 **At the top of the windpipe, making a bulge at the front of the neck, is the voice-box or larynx.** It has two stiff flaps, vocal cords, which stick out from its sides. Normally these flaps are apart for easy breathing. But muscles in the voice-box can pull the flaps almost together. As air passes through the narrow slit between them it makes the flaps shake or vibrate – and this is the sound of your voice.

▼ The vocal cords are held apart for breathing (left) and pulled together for speech (right).

HUMMMMMM!

You will need:

A stopwatch

Do you think making sounds with your voice-box uses more air than breathing? Find out by following this experiment.

1. Take a deep breath in, then breathe out at your normal rate, for as long as you can. Time the out-breath.

2. Take a similar deep breath in, then hum as you breathe out, again for as long as you can. Time the hum.

3. Try the same while whispering your favourite song, then again when singing.

Breathing bits

47 The main parts of the respiratory (breathing) system are the two lungs in the chest. Each one is shaped like a tall cone, with the pointed end at shoulder level.

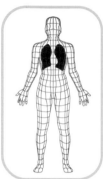

48 Air comes in and out of the lungs along the windpipe, which branches at its base to form two main air tubes, the bronchi. One goes to each lung. Inside the lung, each bronchus divides again and again, becoming narrower each time. Finally the air tubes, thinner than hairs, end at groups of tiny 'bubbles' called alveoli.

Right lung

Left bronchus

Muscles in wall of bronchus

Air space inside bronchus

View along inside of bronchus

49 There are more than 100 million tiny air bubbles, or alveoli, in each lung. Inside, oxygen from breathed-in air passes through the very thin linings of the alveoli to equally tiny blood vessels on the other side. The blood carries the oxygen away, around the body. At the same time a waste substance, carbon dioxide, seeps through the blood vessel, into the alveoli. As you breathe out, the lungs blow out the carbon dioxide.

50 Breathing needs muscle power! The main breathing muscle is the dome-shaped diaphragm at the base of the chest. To breathe in, it becomes flatter, making the lungs bigger, so they suck in air down the windpipe. At the same time, rib muscles lift the ribs, also making the lungs bigger.

To breathe out, the diaphragm and rib muscles relax. The stretched lungs spring back to their smaller size and blow out stale air.

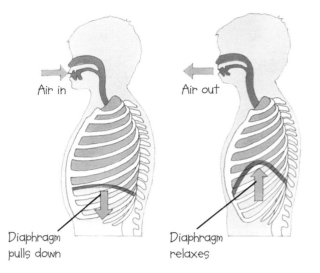

Air in

Air out

Diaphragm pulls down

Diaphragm relaxes

▲ Breathing uses two main sets of muscles, the diaphragm and those between the ribs.

▶ After great activity, the body breathes faster and deeper, to replace the oxygen used by the muscles for energy.

Bronchiole

Blood vessel

Air space in alveoli

Alveoli

▲ Inside each lung, the main bronchus divides again and again, into thousands of narrower airways called bronchioles.

51 As you rest or sleep, each breath sends about half a litre of air in and out, 15 to 20 times each minute. After great activity, such as running a race, you need more oxygen. So you take deeper breaths faster – 3 litres or more of air, 50 times or more each minute.

The hungry body

52 All machines need fuel to make them go, and the body is like a living machine whose fuel is food. Food gives us energy for our body processes inside, and for breathing, moving, talking and every other action we make. Food also provides raw materials that the body uses to grow, maintain itself and repair daily wear-and-tear.

▶ Foods such as bread, pasta and rice contain lots of starch, which is a useful energy source.

▶ Fish, low-fat meats like chicken, and dairy produce such as eggs all contain plenty of valuable proteins.

53 We would not put the wrong fuel into a car engine, so we should not put unsuitable foods into the body. A healthy diet needs a wide variety of different foods, especially fresh vegetables and fruits, which have lots of vital nutrients. Too much of one single food may be unhealthy, especially if that food is very fatty or greasy. Too much of all foods is also unhealthy. It makes the body overweight, which increases the risk of various illnesses.

▶ Cheeses, and fatty and oily foods, are needed in moderate amounts. Plant oils are healthier than fats and oils from animal sources.

54 There are six main kinds of nutrients in foods, and the body needs balanced amounts of all of them.

• Proteins are needed for growth and repair, and for strong muscles and other parts.

• Carbohydrates, such as sugars and starches, give plenty of energy.

• Some fats are important for general health and energy.

• Vitamins help the body to fight germs and disease.

• Minerals are needed for strong bones and teeth and also healthy blood.

• Fibre is important for good digestion and to prevent certain bowel disorders.

▲ Fresh fruits such as bananas, and vegetables such as carrots, have lots of vitamins, minerals and fibre, and are good for the body in lots of ways.

FOOD FOR THOUGHT

Which of these meals do you think is healthier?

Meal A
Burger, sausage and lots of chips, followed by ice-cream with cream and chocolate.

Meal B
Chicken, tomato and a few chips, followed by fresh fruit salad with apple, banana, pear and melon.

Answer:
Meal B

Bite, chew, gulp

55 The hardest parts of your whole body are the ones that make holes in your food – teeth. They have a covering of whitish or yellowish enamel, which is stronger than most kinds of rocks! Teeth need to last a lifetime of biting, nibbling, gnashing, munching and chewing. They are your own food processors.

Incisor

Canine

Premolar

Root

Molar

Jaw bone

◀ In an adult, each side (left and right) of each jaw (upper and lower) usually has eight different-shaped teeth, of four main types.

56 There are four main shapes of teeth. The front ones are incisors, and each has a straight, sharp edge, like a spade or chisel, to cut through food. Next are canines, which are taller and more pointed, used mainly for tearing and pulling. Behind them are premolars and molars, which are lower and flatter with small bumps, for crushing and grinding.

▼ At the centre of a tooth is living pulp, with many blood vessels and nerve endings that pass into the jaw bone.

57 A tooth may look almost dead, but it is very much alive. Under the enamel is slightly softer dentine. In the middle of the tooth is the dental pulp. This has blood vessels to nourish the whole tooth, and nerves that feel pressure, heat, cold and pain. The lower part of the tooth, strongly fixed in the jaw bone, is the root. The enamel-covered part above the gum is the crown.

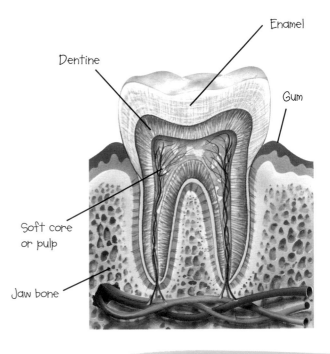

Dentine

Enamel

Gum

Soft core or pulp

Jaw bone

58 Teeth are very strong and tough, but they do need to be cleaned properly and regularly. Germs called bacteria live on old bits of food in the mouth. They make waste products which are acid and eat into the enamel and dentine, causing holes called cavities. Which do you prefer — cleaning your teeth after main meals and before bedtime, or the agony of toothache?

▶ Clean your teeth by brushing in different directions and then flossing between them. They will look better and stay healthier for longer.

▼ The first set of teeth lasts about ten years, while the second set can last ten times longer.

First set
(milk or deciduous teeth)

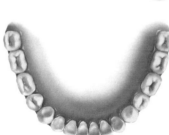

Second set
(adult or permanent set)

59 Teeth are designed to last a lifetime. Well, not quite, because the body has two sets. There are 20 small teeth in the first or baby set. The first ones usually appear above the gum by about six months of age, the last ones at three years old. As you and your mouth grow, the baby teeth fall out from about seven years old. They are replaced by 32 larger teeth in the adult set.

60 After chewing, food is swallowed into the gullet (oesophagus). This pushes the food powerfully down through the chest, past the heart and lungs, into the stomach.

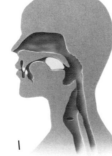
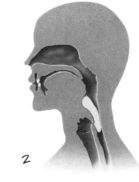

1 tongue pushes food to the back of the throat

2 throat muscles squeeze the food downwards

3 the oesophagus pushes food to the stomach

Food's long journey

61 The digestive system is like a tunnel about 9 metres long, through the body. It includes parts of the body that bite food, chew it, swallow it, churn it up and break it down with natural juices and acids, take in its goodness, and then get rid of the leftovers.

62 The stomach is a bag with strong, muscular walls. It stretches as it fills with food and drink, and its lining makes powerful digestive acids and juices called enzymes, to attack the food. The muscles in its walls squirm and squeeze to mix the food and juices.

63 The stomach digests food for a few hours into a thick mush, which oozes into the small intestine. This is only 4 centimetres wide, but more than 5 metres long. It takes nutrients and useful substances through its lining, into the body.

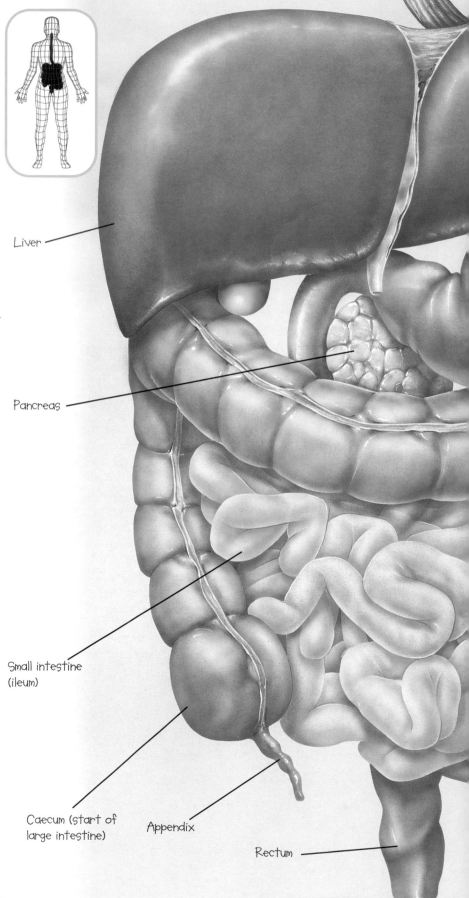

Liver

Pancreas

Small intestine (ileum)

Caecum (start of large intestine)

Appendix

Rectum

64 The large intestine follows the small one, and it is certainly wider, at about 6 centimetres, but much shorter, only 1.5 metres. It takes in fluids and a few more nutrients from the food, and then squashes what's left into brown lumps, ready to leave the body.

Stomach

Large intestine

Vessels inside villus

Villus

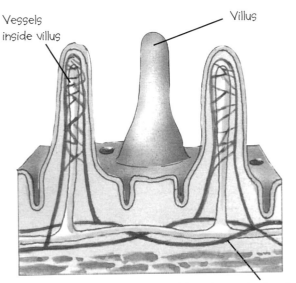

Vessels in intestine lining

▶ The lining of the small intestine has thousands of tiny finger-like parts called the villi, which take nutrients from food, into the blood and lymph system.

◀ The digestive parts almost fill the lower part of the main body, called the abdomen.

65 The liver and pancreas are also parts of the digestive system. The liver sorts out and changes the many nutrients from digestion, and stores some of them. The pancreas makes powerful digestive juices that pass to the small intestine to work on the food there.

I DON'T BELIEVE IT!

What's in the leftovers? The brown lumps called bowel motions or faeces are only about one-half undigested or leftover food. Some of the rest is rubbed-off parts of the stomach and intestine lining. The rest is millions of 'friendly' but dead microbes (bacteria) from the intestine. They help to digest our food for us, and in return we give them a warm, food-filled place to live.

Blood in the body

66 The heart beats to pump the blood all around the body and pass its vital oxygen and nutrients to every part. The same blood goes round and round, or circulates, in its network of blood vessels. So the heart, blood vessels and blood are known as the circulatory system.

Carotid artery

◀ Blood vessels divide, or branch, to reach every body part.

Blood vessels in lung

▶ There are three main kinds of blood vessels.

Capillary

Vein

Artery

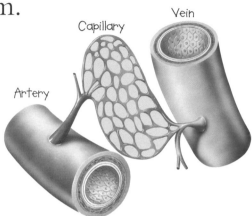

Heart

Iliac artery

67 Blood travels from the heart through strong, thick-walled vessels called arteries. These divide again and again, becoming smaller until they form tiny vessels narrower than hairs, called capillaries. Oxygen and nutrients seep from the blood through the thin capillary walls to the body parts around. At the same time, carbon dioxide and waste substances seep from body parts into the blood, to be carried away. Capillaries join again and again to form wide vessels called veins, which take blood back to the heart.

68 In addition to delivering oxygen and nutrients, and carrying away carbon dioxide and wastes, blood has many other vital tasks. It carries body control substances called hormones (see page 43). It spreads heat evenly around the body from busy, warmer parts such as the heart, liver and muscles. It forms a sticky clot to seal a cut. It carries many substances that attack germs and other tiny invaders.

69
Blood has four main parts. The largest is billions of tiny, saucer-shaped red cells, which make up almost half of the total volume of blood and carry oxygen. Second is the white cells, which clean the blood, prevent disease and fight germs. The third part is billions of tiny platelets, which help blood to clot. Fourth is watery plasma, in which the other parts float.

Muscle layer
Elastic layer
Tough outer cover
Inner lining
Plasma
Red cell
White cell
Platelet

▼ A blood vessel wall has several layers, and blood itself contains different types of cells.

QUIZ
Can you match these blood parts and vessels with their descriptions?
a. Artery b. Vein c. White blood cell
d. Red blood cell e. Platelet f. Capillary

1. Large vessel that takes blood back to the heart
2. Tiny vessel allowing oxygen and nutrients to leave blood
3. Large vessel carrying blood away from the heart
4. Oxygen–carrying part of the blood
5. Disease–fighting part of the blood
6. Part that helps blood to clot

Answers:
a3 b1 c5 d4 e6 f2

70
Blood is cleaned by two kidneys, situated in the middle of your back. They filter the blood and make a liquid called urine, which contains unwanted and waste substances, plus excess or 'spare' water. The urine trickles from each kidney down a tube, the ureter, into a stretchy bag, the bladder. It's stored here until you can get rid of it – at your convenience.

Cortex
Medulla
Blood vessels
Ureter

▲ Each kidney has about one million tiny filters, called nephrons, in its outer layer, or cortex.

The beating body

71 The heart is about as big as its owner's clenched fist. It is a hollow bag of very strong muscle, called cardiac muscle or myocardium. This muscle never tires. It contracts once every second or more often, all through life. The contraction, or heartbeat, squeezes blood inside the heart out into the arteries. As the heart relaxes it fills again with blood from the veins.

72 Inside, the heart is not one baglike pump, but two pumps side by side. The left pump sends blood all around the body, from head to toe, to deliver its oxygen (systemic circulation). The blood comes back to the right pump and is sent to the lungs, to collect more oxygen (pulmonary circulation). The blood returns to the left pump and starts the whole journey again.

► The heart is two pumps side by side, and each pump has two chambers, the upper atrium and the lower ventricle.

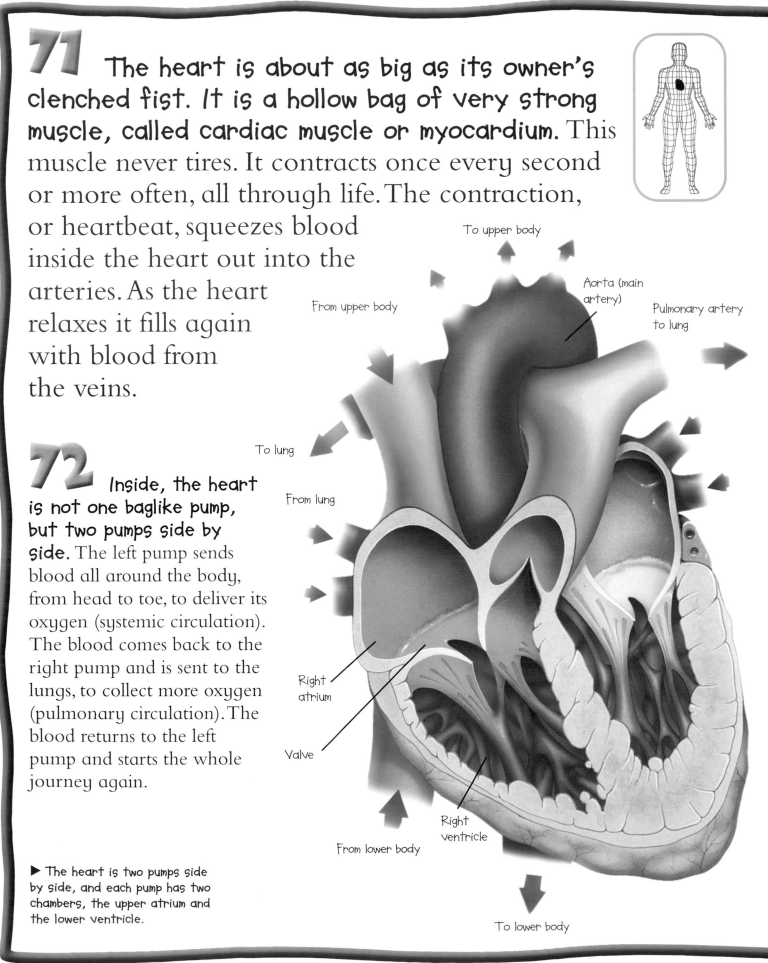

To upper body

From upper body

Aorta (main artery)

Pulmonary artery to lung

To lung

From lung

Right atrium

Valve

Right ventricle

From lower body

To lower body

73 Inside the heart are four sets of bendy flaps called valves. These open to let blood flow the right way. If the blood tries to move the wrong way, it pushes the flaps together and the valve closes. Valves make sure the blood flows the correct way, rather than sloshing to and fro, in and out of the heart, with each beat.

▶ The heartbeat is the regular squeezing of the heart muscle to pump blood around the body.

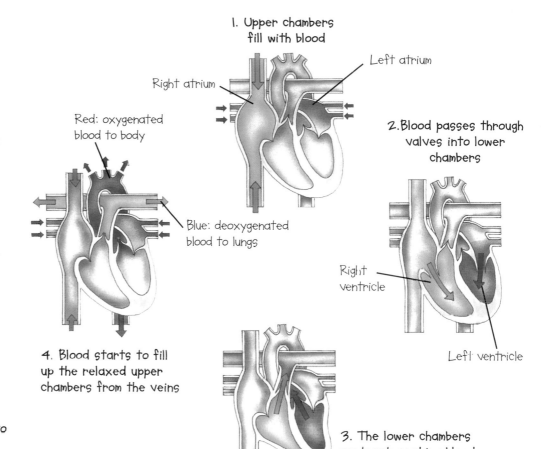

1. Upper chambers fill with blood

Left atrium

Right atrium

Red: oxygenated blood to body

Blue: deoxygenated blood to lungs

2. Blood passes through valves into lower chambers

Right ventricle

Left ventricle

4. Blood starts to fill up the relaxed upper chambers from the veins

3. The lower chambers contract, pushing blood into the arteries

74 The heart is the body's most active part, and it needs plenty of energy brought by the blood. The blood flows through small vessels, which branch across its surface and down into its thick walls. These are called the coronary vessels.

75 The heart beats at different rates, depending on what the body is doing. When the muscles are active they need more energy and oxygen, brought by the blood. So the heart beats faster, 120 times each minute or more. At rest, the heart slows to 60 to 80 beats per minute.

HOW FAST IS YOUR HEARTBEAT?

You will need:

plastic funnel   tracing paper
plastic tube (like hosepipe)   sticky-tape

You can hear your heart and count its beats with a sound-funnel device called a stethoscope.

1. Stretch the tracing paper over the funnel's wide end and tape in place. Push a short length of tube over the funnel's narrow end.

2. Place the funnel's wide end over your heart, on your chest, just to the left, and put the tube end to your ear. Listen to and count your heartbeat.

Looking and listening

76 The body finds out about the world around it by its senses – and the main sense is eyesight. The eyes detect the brightness, colours and patterns of light rays, and change these into patterns of nerve signals that they send to the brain. More than half of the knowledge, information and memories stored in the brain come into the body through the eyes.

▶ The eye is moved by six tiny muscles, and inside, it is filled with a clear fluid, vitreous humour.

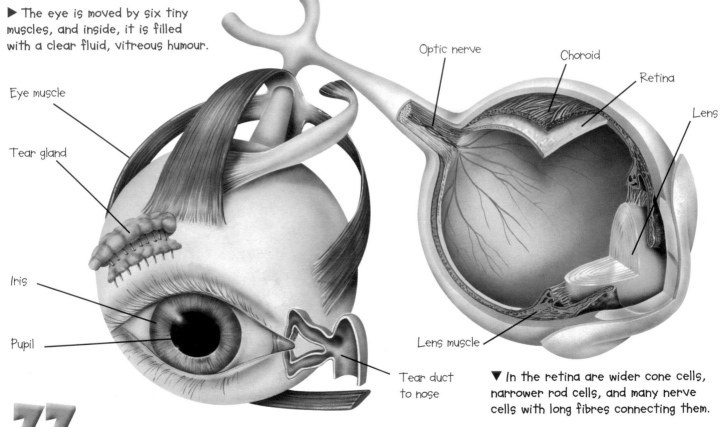

Eye muscle

Tear gland

Iris

Pupil

Optic nerve

Choroid

Retina

Lens

Lens muscle

Tear duct to nose

▼ In the retina are wider cone cells, narrower rod cells, and many nerve cells with long fibres connecting them.

Rod cell

Cone cell

Nerve cells

77 Each eye is a ball about 2.5 centimetres across. At the front is a clear dome, the cornea, which lets light through a small, dark-looking hole just behind it, the pupil. The light then passes through a pea-shaped lens which bends the rays so they shine a clear picture onto the inside back of the eye, the retina. This has 125 million tiny cells, rods and cones, which detect the light and make nerve signals to send along the optic nerve to the brain.

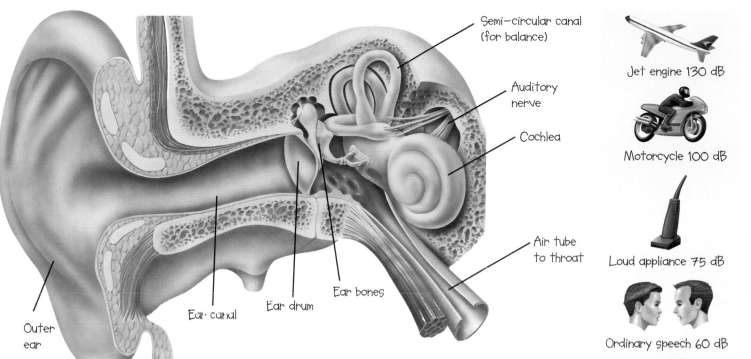

Semi-circular canal (for balance)

Auditory nerve

Cochlea

Air tube to throat

Ear bones

Ear drum

Ear canal

Outer ear

▲ Most of the small, delicate parts of the ear are inside the head, well protected by skull bones around them.

Jet engine 130 dB

Motorcycle 100 dB

Loud appliance 75 dB

Ordinary speech 60 dB

Whisper 20 dB

▶ The loudness, or volume, of sounds is measured in decibels (dB). Louder than about 90dB can damage hearing.

78

The ear is far more than the bendy, curly flap on the side of the head. The ear flap funnels sound waves along a short tunnel, the ear canal, to a fingernail-sized patch of tight skin, the eardrum. As sound waves hit the eardrum it shakes or vibrates, and passes the vibrations to a row of three tiny bones. These are the ear ossicles, the smallest bones in the body. They also vibrate and pass on the vibrations to another part, the cochlea, which has a curly, snail-like shape.

BRIGHT AND DIM

Look at your eyes in a mirror. See how the dark hole which lets in light, the pupil, is quite small. The coloured part around the pupil, the iris, is a ring of muscle.

Close your eyes for a minute, then open them and look carefully. Does the pupil quickly get smaller?

While the eyes were closed, the iris made the pupil bigger, to try and let in more light, so you could try to see in the darkness. As you open your eyes, the iris makes the pupil smaller again, to prevent too much light from dazzling you.

79

Inside the cochlea, the vibrations pass through fluid and shake rows of thousands of tiny hairs which grow from specialized hair cells. As the hairs vibrate, the hair cells make nerve signals, which flash along the auditory nerve to the brain.

Smelling and tasting

Olfactory cells

Mucus lining

Nasal cavity

80 You cannot see smells, which are tiny particles floating in the air – but your nose can smell them. Your nose is more sensitive than you realize. It can detect more than 10,000 different scents, odours, perfumes, fragrances, pongs and niffs. Smell is useful because it warns us if food is bad or rotten, and perhaps dangerous to eat. That's why we sniff a new or strange food item, almost without thinking, before trying it.

▼ Olfactory (smell) cells have micro-hairs facing down into the nasal chamber, which detect smell particles landing on them.

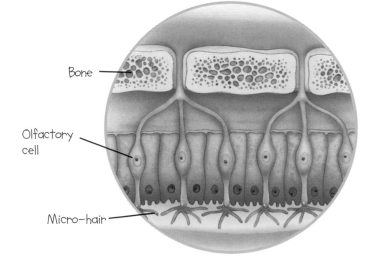

Bone

Olfactory cell

Micro-hair

81 Smell particles drift with breathed-in air into the nose and through the nasal chamber behind it. At the top of the chamber are two patches of lining, each about the area of a thumbnail and with 250 million microscopic hairs. The particles land on the sticky hairs, and if they fit into landing sites called receptors there, like a key into a lock, then nerve signals flash along the olfactory nerve to the brain.

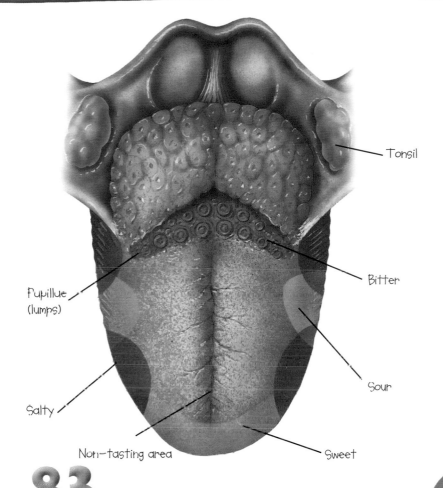

- Tonsil
- Bitter
- Sour
- Sweet
- Non-tasting area
- Salty
- Papillae (lumps)

The body's most flexible muscle is also the one which is coated with 10,000 micro-sensors for taste – the tongue. Each micro-sensor is a taste bud shaped like a tiny onion. Most taste buds are along the tip, sides and rear upper surface of the tongue. They are scattered around the much larger flaps and lumps on the tongue, which are called papillae.

◄ The taste buds are mainly around the edges of the tongue, not on the main middle area.

83 Taste works in a similar way to smell, but it detects flavour particles in foods and drinks. The particles touch tiny hairs sticking up from hair cells in the taste buds. If the particles fit into receptors there, then the hair cell makes nerve signals, which go along the facial and other nerves to the brain.

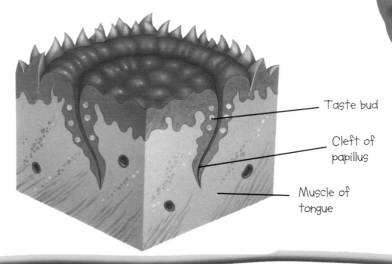

- Taste bud
- Cleft of papillus
- Muscle of tongue

SWEET AND SOUR

The tongue detects only four basic flavours – sweet at the tip, salty along the front sides, sour along the rear sides, and bitter across the back.

Which of these foods is sweet, salty, bitter or sour?

1. Coffee 2. Lemon 3. Bacon
4. Ice cream

Answers:
1. bitter 2. sour 3. salty 4. sweet

◄ The large pimple-like lumps at the back of the tongue, called papillae, have tiny taste buds in their deep clefts.

The nervous body

Brain

Spinal cord

Sciatic nerve

Tibial nerve

84 **The body is not quite a 'bag of nerves', but it does contain thousands of kilometres of these pale, shiny threads.** Nerves carry tiny electrical pulses known as nerve signals or neural messages. They form a vast information-sending network that reaches every part, almost like the body's own Internet.

85 **Each nerve is a bundle of much thinner parts called nerve fibres.** Like wires in a telephone cable, these carry their own tiny electrical nerve signals. A typical nerve signal has a strength of 0.1 volts (one-fifteenth as strong as a torch battery). The slowest nerve signals travel about half a metre each second, the fastest at more than 100 metres per second.

Axon

Dendrites

▲ Nerves branch from the brain and spinal cord to every body part.

86 **All nerve signals are similar, but there are two main kinds, depending on where they are going.** Sensory nerve signals travel from the sensory parts (eyes, ears, nose, tongue and skin) to the brain. Motor nerve signals travel from the brain out to the muscles, to make the body move about.

Synapse (junction between nerve cells)

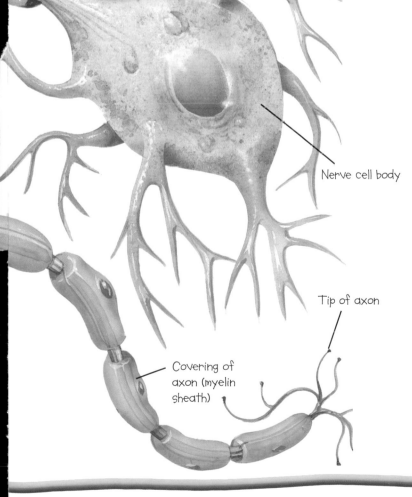

TIME TO REACT!
You will need:
a friend   ruler

1. Ask your friend holds the ruler by the end with the highest measurement, letting it hang down. Put your thumb and fingers level with the other end, ready to grab.

2. Get your friend to let the ruler go, for you to grasp it as it falls. Measure where your thumb is on the ruler. Swap places so your friend has a go.

3. The person who grabs the ruler nearest its lower end has the fastest reactions. To grab the ruler, nerve signals travel from the eye, to the brain, and back out to the muscles in the arm and hand.

Nerve cell body

Tip of axon

Covering of axon (myelin sheath)

87 Hormones are part of the body's inner control system. A hormone is a chemical made by a gland. It travels in the blood and affects other body parts, for example, making them work faster or release more of their product.

▼ Female and male bodies have much the same hormone–making glands, except for the reproductive parts — ovaries in the female (left) and testes in the male (right).

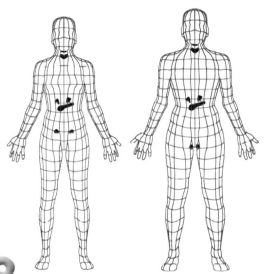

88 The main hormonal gland, the pituitary, is also the smallest. Just under the brain, it has close links with the nervous system. It mainly controls other hormonal glands. One is the thyroid in the neck, which affects the body's growth and how fast its chemical processes work. The pancreas controls how the body uses energy, by its hormone, insulin. The adrenal glands are involved in the body's balance of water, minerals and salts, and how we react to stress and fear.

◄ The brain and nerves are made of billions of specialized cells, nerve cells or neurons. Each has many tiny branches, dendrites, to collect nerve messages, and a longer, thicker branch, the axon or fibre, to pass on the messages.

The brainy body

89 Your brain is as big as your two fists side by side. It's the place where you think, learn, work out problems, remember, feel happy and sad, wonder, worry, have ideas, sleep and dream.

▼ The two wrinkled hemispheres (halves) of the cerebrum, where thinking happens, are the largest brain parts.

Cerebral hemisphere

Thalamus

Hippocampus

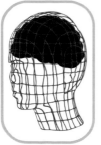

Cerebellum

Brain stem

91 The main part of the brain is its bulging, wrinkled upper part, the cerebrum. Different areas of its surface (cerebral cortex) deal with nerve signals to and from different parts of the body. For example, messages from the eyes pass to the lower rear part of the cerebrum, called the visual centre. They are sorted here as the brain cells work out what the eyes are seeing. There are also areas for touch, hearing, taste and other body processes.

90 Yet the brain looks like a wrinkly lump of grey–pink jelly! On average, it weighs about 1.4 kilograms. It doesn't move, but its amazing nerve activity uses up one-fifth of all the energy needed by the body.

92

The cerebellum is the rounded, wrinkled part at the back of the brain. It processes messages from the motor centre, sorting and coordinating them in great detail, to send to the body's hundreds of muscles. This is how we learn skilled, precise movements such as writing, skateboarding or playing music (or all three), almost without thinking.

93

The brain stem is the lower part of the brain, where it joins the body's main nerve, the spinal cord. The brain stem controls basic processes vital for life, like breathing, heartbeat, digesting food and removing wastes.

94

The brain really does have 'brain waves'. Every second it receives, sorts and sends millions of nerve signals. Special pads attached to the head can detect these tiny electrical pulses. They are shown on a screen or paper strip as wavy lines called an EEG, electro-encephalogram.

▼ Different areas or centres of the brain's outer layer, the cerebral cortex, deal with messages from and to certain parts of the body.

Touch area
Movement area
Thought area
Vision area
Hearing area
Speech area

▼ The brain's 'waves' or EEG recordings change, depending on whether the person is alert and thinking hard, resting, falling asleep or deeply asleep.

I DON'T BELIEVE IT!

The brain never sleeps! EEG waves show that it is almost as busy at night as when we are awake. It still controls heartbeat, breathing and digestion. It also sifts through the day's events and stores memories.

The healthy body

95 No one wants to be ill – and it is very easy to cut down the risk of becoming sick or developing disease. For a start, the body needs the right amounts of different foods, especially fresh foods like vegetables and fruits. And not too much food either, or it becomes unhealthily fat.

96 Another excellent way to stay well is regular sport or exercise. Activity keeps the muscles powerful, the bones strong and the joints flexible. If it speeds up your breathing and heartbeat, it keeps your lungs and heart healthy too.

◀ Germs on hands can get onto our food and then into our bodies. So it is important to wash hands before mealtimes.

97 Germs are everywhere – in the air, on our bodies and on almost everything we touch. If we keep clean by showering or bathing, and especially if we wash our hands after using the toilet and before eating, then germs have less chance to attack us.

98 Health is not only in the body, it's in the mind. Too much worry and stress can cause many illnesses, such as headaches and digestive upsets. This is why it's so important to talk about troubles and share them with someone who can help.

◄ Exercise keeps the body fit and healthy, and it should be fun too. It is always best to reduce risks of having an accident by wearing a cycle helmet for example.

► In some immunizations, dead versions of a germ are put into the body using a syringe, so the body can develop resistance to them without suffering from the disease they cause.

99 Doctors and nurses help us to recover from sickness, and they also help prevent illness. Regular check-ups at the dentist, optician and health centre are vital. For most people immunizations (vaccinations) also help to protect against diseases. It is good to report any health problem early, before they become too serious to treat.

100 Old age is getting older! More people live to be 100 years or more and for many of them, their bodies are still working well. How would you like to spend your 100th birthday?

Index